HORRIBLE SCIENCE

TEACHERS' RESOURCES

FORCES

Nick Arnold • Tony De Saulles
additional material David Tomlinson

AUTHOR
Nick Arnold

ILLUSTRATIONS
Tony De Saulles

ADDITIONAL MATERIAL
David Tomlinson

EDITOR
Wendy Tse

ASSISTANT EDITOR
Charlotte Ronalds

SERIES DESIGNER
Joy Monkhouse

DESIGNER
Erik Ivens

This book contains extracts from *Fatal Forces*, *The Awfully Big Quiz Book* and *Explosive Experiments* in the Horrible Science series. Text © 1997, 2000, 2001, Nick Arnold. Illustrations © 1997, 2000, 2001, Tony De Saulles. First published by Scholastic Children's Books. Additional text © 2004, David Tomlinson.

Designed using Adobe InDesign

Published by Scholastic Ltd
Villiers House
Clarendon Avenue
Leamington Spa
Warwickshire
CV32 5PR

www.scholastic.co.uk

Printed by Bell & Bain Ltd, Glasgow

4 5 6 7 8 9   5 6 7 8 9 0 1 2 3

British Library Cataloguing-in-Publication Data
A catalogue record for this book is available from the British Library.

ISBN 0-439-97182-9
The right of David Tomlinson to be identified as the Author of additional text of this Work has been asserted by him in accordance with the Copyright, Designs and Patents Act 1988.

TEACHERS' NOTES

Horrible Science Teachers' Resources: Forces is inspired by the Horrible Science book *Fatal Forces*. Each photocopiable page takes a weird and wonderful excerpt from the original, as well as material from *Explosive Experiments* and *The Awfully Big Quiz Book*, and expands on it to create a class-based teaching activity, fulfilling both National Curriculum and QCA objectives. The activities can be used individually or in a series as part of your scheme of work.

With an emphasis on research, experimentation and interpreting results, the activities will appeal to anyone even remotely curious about the Horrible world around us!

PART 1: INTRODUCING FORCES

Page 11: Introducing Isaac
Learning objective
Objects are pulled downwards by gravity.
Earth and objects are pulled together.

Start the session by giving one of your class an apple and tell them that you will soon ask them to drop it. Before they drop it, ask your class what they think will happen, and ask them to explain why. Once the child has dropped the apple introduce the word 'gravity', using photocopiable page 11 to focus your class on the story of Isaac Newton and the study of forces.
Answers: All are true except **2)** Newton never married.

Page 12: Newton's First Law
Learning objective
Identifying the direction of forces.
Objects can move quicker and faster.

Using the target on photocopiable page 12, encourage your class to explore Newton's First Law and to apply it to situations around them. Once the children have played the game, challenge them to design other games that use this law and to write a 'Games Guide', explaining how the First Law is applied. They can use these to compile a class Gravity Games book.

Page 13: Newton's Second Law
Learning objective
Identifying the direction of forces.
Objects can move quicker and faster.

Start this session by asking your children if they have ever played tabletop football, using players that are flicked as opposed to being attached to a pole. Apply Newton's Second Law to this type of game and use photocopiable page 13 to set up a class tournament. Encourage your class to think through the actions that they take on the 'pitch' and to write a match commentary, highlighting the role that Newton's Second Law plays.

Page 14: Newton's Third Law
Learning objective
Measuring forces.
Using results to draw conclusions.

Use photocopiable page 14 to introduce the concept that although we cannot always see forces at work, they are still around us. Forces work in pairs and for every action there is an equal reaction. Use examples from everyday life (for example, a punch bag) and explain that we can gauge how strong a force is by the reaction to it. In this experiment gravity pushes down on the book and the reactive force of the table pushes back. This stops the book from falling to the floor. As the reactive force from the table is an equal force it means that the table does not push the book into the air! If the book is heavy enough, the effect of the equal and opposite forces can be seen in the changes to the Plasticine.

Recap any experience the children may have had of making cookies or Plasticine modelling, explaining that they must look closely at the indents made by the cutter to be able to tell the pressure of the force that was exerted upon it. Use the results for a class display.

Page 15: Newtons
Learning objective
How to measure forces.
Measuring in Newtons.

Start this session by showing the children a variety of measuring equipment and ask them to sort the equipment according to the job they do (for example, measuring liquid by volume, length in cm and metres, weight in kg, and so on). Explain that we use the word 'weight' to mean both *mass* and *weight* but in scientific terms there is a difference between the weight of an object and its mass. A Newton meter measures the effect of the pull of gravity on an object. The unit of measurement is 'Newtons' (N). Encourage your class to estimate weights of objects in Newtons, using photocopiable page 15 to record their results.
Answer: c)

Page 16: Measuring on the moon!
Learning objective
How to measure forces.
Measuring in Newtons.

Recap any work you may have done about the effects of gravity on Earth, explaining that zero gravity is experienced in space and reduced gravity is experienced on the moon. Introduce the children to a range of objects from your classroom, explaining that you are all going to have the opportunity to weigh them here on Earth. Make sure that all the scales are set at zero and encourage your class to work out why it is important not to lean on an object while it is being weighed.

Use photocopiable page 16 as the starting point for division calculations to explore the weight of earth-bound objects in space, applying any division strategies you have been using in recent maths lessons.

PART 2:
INVESTIGATING FORCES

Page 17: Idle inertia
Learning objective
Identifying the direction of forces.
Descriptions and explanations involving a sequence of ideas.

Start this session by sitting very still, asking the children if you are moving. Explain that although you may appear to be still, your heart is still beating, your lungs are still breathing and so there is hidden movement. Use photocopiable page 17 to relate this concept to 'inertia', encouraging the children to write up the egg experiment in their own words.
Answer: b) When you stop the raw egg, inertia keeps the egg white inside spinning. And this starts the entire egg spinning again when you lift your finger. The inside of the hard-boiled egg is hard, of course, so the white doesn't have its own inertia.

Page 18: Momentum
Learning objective
Identifying the direction of forces.
Descriptions and explanations involving a sequence of ideas.

Recap any work your class may have done on inertia and contrast this to the examples on photocopiable page 18. Explain that a person or object's mass and speed combine to produce 'momentum'. Encourage the children to start a 'Momentum File', drawing examples from the classroom and home life to use for a class presentation (for example, riding a bike, pushing a computer trolley, rolling a marble).

Pages 19 & 20: Galileo's gravity 1 & 2
Learning objective
Objects are pulled downwards by gravity.
Earth and objects are pulled together.

Recap any work you may have done about gravity, reminding the children of examples they have identified. Use photocopiable pages 19 and 20 to show how early scientists undertook the scientific process. Encourage the children to suggest their own ideas, stressing that Galileo and other significant scientists often reach their conclusions through trial and error.

A fun but messier alternative to the experiment on photocopiable page 19 is to use two balloons of the same size: one filled with air, the other with water. The children can drop them, while standing on a chair, into a basin on the floor. This is not recommended for carpeted floors!

For the experiment on photocopiable page 20, you could use plastic guttering or the cardboard tubes from wrapping paper for the long, wide tube. **Answers:** In both cases, Galileo's experiments and conclusions were correct.

Page 21: Friction
Learning objective
Friction as a force that slows moving objects. Force between two moving surfaces is friction.

Start the session with the hand-rubbing activity suggested on photocopiable page 21. Ask the children to predict what will happen and to explain their reasoning, contrasting these ideas with the actual results. Stress that this is how all experiments are carried out and introduce the ideas of fair testing and hypothesis. Relate the main activity to situations around us: for example, the grip of shoes on different surfaces. Encourage the children to write up the experiment in their own words and to compare results.

THESE 'FRICTION' BOOTS ARE GREAT FOR WALKING UP STEEP HILLS!

Page 22: Avoiding friction!
Learning objective
Friction as a force that slows moving objects. Force between two moving surfaces is friction.

Recap any work you may have done about friction, contrasting situations where it is useful and others where it can be dangerous. Use photocopiable page 22 to focus your class on ways that friction can be avoided and encourage them to give reasons for their ideas.
SAFETY NOTE! Ensure the children do not eat any of the foodstuffs used in this activity.
Answer: a) Lubricating oils are squeezed from peanuts, coconuts or bits of dead fish. In some countries bananas are used because they're slippery too. That's why you slip on a banana skin!

Page 23: Stretching
Learning objective
Thinking creatively in science.
Asking scientific questions.

Start by asking the children to stretch out, encouraging them to describe what they see and feel. Use photocopiable page 23 to focus the children on what 'stretching' means scientifically, asking them for examples from everyday life. Use the activities to highlight the link between weight and stretch.
SAFETY NOTE! Remind your class that they must NEVER flick elastic bands at each other as this can be dangerous, and that the bands must not go into their mouths under any circumstances.
'What happens when something stretches?'
answer: b) The band briefly stores energy from the force that stretches it. The energy tries to escape as heat and that's why the band feels hot.
'The power of an elastic band?' answer: b) The machine uses the force you put into turning the elastic band to move. The friction provided by rough slopes helps your machine grip the ground and climb better.

Page 24: Springs around us!
Learning objective
Thinking creatively in science.
Asking scientific questions.

Encourage your class to put together a collection of springs, a particularly useful one is a 'Slinky' toy, which will illustrate how springs can move and generate movement in other objects. Use photocopiable page 24 to focus the children on different springs, encouraging them to use this knowledge to make their own Jack in the Box, complete with paper concertina spring. Stiff paper or thin cardboard is best for this activity. Show the Jack in the Boxes to visiting classes, encouraging your children to explain how they were made.

Page 25: Centrifugal effect and centripetal force
Learning objective
Considering sources of information.
Asking scientific questions.

Use the explanation box on photocopiable page 25 to explain the basic theory of centrifugal effect and centripetal force, asking the children to relate any examples that they think are to do with these forces. Stress that while this theory is overall referred to as two 'forces', centrifugal force is actually an effect rather than a force. It is an example of Newton's Third Law: when an object exerts a force on another object, the second object will push back just as hard.

Use a chair outside as the target for this activity, rewarding the class's observations of the force and the effect. Encourage the children to write up the experiment, and remind them to include their own hypotheses.
Answer: b) When you release the string, centrifugal force makes the bolas fly off at high speed in a straight line. When the string hits the tree centripetal force on the string pulls the balls inwards so they wrap round the trunk.

Page 26: The centrifugal effect
Learning objective
Thinking creatively in science.
Making generalisations.

Explain to your class that the activity on photocopiable page 26 is based on a game played at Isaac Newton's school. It may even have helped him to unravel the mystery of gravity and the centrifugal effect.
SAFETY NOTE! Ensure that the buckets are firmly attached and that the playground is empty before attempting! Encourage the children to write up the experiment using the appropriate scientific headings.

Page 27: Spinning
Learning objective
Thinking creatively in science.
Asking scientific questions.

Start this session by asking your class if they have used spinning tops (marketed under different brand names), explaining that these toys are amongst the oldest in the world and that they have their roots in serious scientific research. Spin a top, asking the children to identify forces they think may be important. Play the Igloo game on photocopiable page 27 and encourage the children to hold a spinning-top tournament. As an extension, use stopwatches to time the class spinning champ! Use the data for a class graph.

Page 28: Push, pull and twist
Learning objective
Pushes and pulls are opposing forces.
Words relating to movement.

Start this session by asking a child to go and open the door, asking him or her to identify each movement (you twist the handle, pull the door, push the door to close it again). Explain that these actions are all forces and we use them every day, often without thinking about it. Ask the children for their own examples, relating them to photocopiable page 28. Encourage the children to draw their own push, pull and twist examples on the blank cards and to mix the cards for games of 'Snap' in pairs or small groups. Explain that when two cards of the same force come up then they should shout 'snap', as in the usual game.

PART 3:
AIR AND WATER

Page 29: Air and water
Learning objective
Air resistance can slow objects.
Resistance can slow an object.

Start this session by showing the children an 'empty' box, asking them to describe what is in it. Explain that although they cannot see it, it is full of 'air' (a mixture that includes oxygen). Introduce the concept of a vacuum, asking for other examples (cleaner, flask and so on). Use photocopiable page 29 to focus the children on the concept of 'full' and 'empty', explaining that although the classroom may be emptied of furniture it is still 'full' of gases, the main two being nitrogen and oxygen. Expand this debate to include everyday vacuums we create and use.
Answer: b) Before you drink you breathe in. This lowers the air pressure in your mouth. The higher air pressure inside the bottle forces the drink into your mouth.

By covering the mouth of the bottle you make the air pressure in the bottle the same as your mouth. The drink won't flow. Don't suck too hard or you might swallow the bottle instead.

Page 30: Air pressure 1
Learning objective
Air resistance can slow objects.
Resistance can slow an object.

Recap any previous work on air pressure, relating it to aeroplanes and everyday life. Show the children a barometer if you have one and explain that it is used to measure air pressure just as a ruler is used to measure length or a Newton meter measures forces. Air pressure gets lower as we go higher, which is why planes have pressurised cabins replicating the air pressure on the ground.
   As we go lower, the pressure gets higher. This is why submarines also need to replicate pressure on the ground to counteract the effect of the water pressure. Divers need to be especially careful when surfacing as they can suffer from the 'bends', which is caused by experiencing the change between water and air pressure too quickly.
   Refer to the 'Bet you never knew' box on photocopiable page 30 to explain how air pressure affects people in aeroplanes. Use the activity to show your class how it occurs.

Page 31: Air pressure 2
Learning objective
Air resistance can slow objects.
Resistance can slow an object.

Use photocopiable page 31 to encourage your class to work in pairs, testing their theories about air pressure and recording their ideas at each opportunity. Explain that air pressure is measured in 'Pascals', named after the French scientist Blaise Pascal who invented the barometer. 1 Pascal = 1 Newton per square metre. Encourage all the children to participate in formulating hypotheses and compare them at the end of the session.

Pages 32 & 33: Air resistance & Air resistance report
Learning objective
Air resistance can slow objects.
Resistance can slow an object.

Start this session by talking about skydiving and parachutists, using pictures from class books. Ask your class why a parachute might be useful when falling through air, focusing on its function as an air resistor. Use photocopiable page 32 to make two parachutes of different sizes, and compare the results of each. Encourage your class to record this experiment on photocopiable page 33, focusing on fair testing. As an extension, ask the children what difference it would make to their results if they cut the cartoon circles out of the parachutes.

PART 4:
FORCES IN ACTION

Page 34: Juggling
Learning objective
Gravitational pull and opposing forces.
Balanced and unbalanced forces.

Start this session by recapping any work on gravity, momentum and inertia that you may have done previously. Combine these forces for the main activity on photocopiable page 34, encouraging the children to give their best effort regardless of natural co-ordination. If the children are confident of their juggling, the class or a group can give a final presentation with music to other classes.

Page 35: Tennis
Learning objective
Gravitational pull and opposing forces.
Balanced and unbalanced forces.

Start this session by explaining that when scientists draw diagrams they use arrows to show the direction of a particular force. Ask the children to draw themselves sitting in their chair, complete

with arrows showing the different forces at work. Use photocopiable page 35 to focus your class on the tennis challenge, explaining that they will be finding out which forces they use. Take your class outside for the activity, encouraging them to draw detailed diagrams, complete with arrows, and captions afterwards. Alternatively, they can try the activity with an indoor tennis set.

Page 36: Bouncing
Learning objective
Gravitational pull and opposing forces.
Balanced and unbalanced forces.

Start this session by asking for a volunteer to bounce a ball at the front of the classroom. Encourage the children to talk about what is happening to the ball, to trace its journey during the bounce and to apply the names of any forces that they think might have something to do with it. Explain that some materials bounce more than others, demonstrating this with a less bouncy object (for example, a paper ball). Link this to any work you may have done about molecules making up the structure of all materials. The structure is different for each material, which is why some materials bounce more than others. Use photocopiable page 36 to focus your class in more detail on how a ball might bounce, applying the concept of fair testing. Encourage the children to think of other materials they could use for this experiment and to try them out, reporting their results.

Page 37: Vibration
Learning objective
Measuring forces.
Checking measurements.

Start the session by talking about vibration around us, using household examples and relating this to air flights and on a larger scale, earthquakes. Use the experiments to illustrate how vibrations can affect us, encouraging the children to write up their findings using scientific headings.
'Dare you discover' answer: b) Your body vibrates constantly as your heart beats and the blood squirts

around your body. Your muscles also pulse of their own accord. So your body's vibrations pass along the ruler, and make it twitch. If you get answer **a)** try a heavier weight. If you get answer **c)** try a lighter one.

Weird whine glasses answer: You hear a ringing sound. This fascinating scientific sound effect is caused by vibrations made by your finger on the glass, but these grow stronger as the glass and the air in the glass vibrates too. Half-fill a glass with water and you'll find that deeper water causes lower sounds. The water slows the speed of the vibrations and makes the sound deeper to your ears.

Page 38: Testing balance
Learning objective
Objects are pulled downwards by gravity.
Using results to draw conclusions.

Start this practical session by talking about situations when it is important for us to keep our balance, encouraging your class to relate their experiences playing sports. Use photocopiable page 38 to record the children's ideas about balancing on a walking beam. Ensure that the beam is no more than 20cm from the floor, as part of the activity involves imbalance and correcting this imbalance. The children are likely to balance using their arms to steady themselves. This is because our weight is redistributed as we move and our arms are compensating. Encourage the children to explain how they kept their test fair. Split your class into groups to look at the statements in the quiz, grouping them as 'likely' and 'unlikely' to be true, giving reasons for their answers.

Answers: 1 TRUE. Some people will do anything to get attention. **2** TRUE. Blondin also went across on stilts. By then he was just showing off. **3** FALSE. **4** TRUE. **5** TRUE. **6** TRUE. He built the record house of cards in Copenhagen, Denmark. Couldn't he find anything more exciting to do? **7** FALSE. In fact he cycled for an incredible 640 minutes.

PART 5:
SCIENCE SKILLS ASSESSMENT CAROUSEL

The five photocopiable sheets in this section are designed for assessment, allowing your children to show off their scientific skills through a series of short practical activities. *They are not intended to provide assessment of factual knowledge or results.* The sheets can be used in a single session. Prepare for this session by ensuring each child has a copy of photocopiable pages 39 to 43 and that the activities are set up and ready around the room. Alternatively, the activities may be used singly with individuals or pairs in normal class conditions.

The carousel comprises:

39: The flying ball test (air pressure)
40: Air power (air and water pressure)
41: Water wonder (water pressure)
42: Paper power (weight)
43: Body balance test (gravity)

Learning objective
Thinking creatively to try to explain how things work, establishing links between cause and effect. Enquiry in environmental and technological contexts.

Start the session by explaining to your class that they will be working in teams on a series of ten-minute tasks. Talk through each of the activities on photocopiable pages 39 to 43 briefly, and explain that each group will move round the classroom like a carousel from one activity to the next. It is a good idea to put down a polythene sheet for the Water wonder experiment (or conduct it outside) because water may be spilt on the floor if it is not done properly.

Clipboards may be useful for keeping the photocopiable sheets together, alternatively staple them into a booklet. Encourage the children to record any extra notes on the back of the sheets. Assess the children by observation and questioning, focusing on the concepts of fair testing, reasoning, hypothesising and understanding their results. Finish with a class award ceremony to celebrate their achievements.

PART 6:
QUIZ

Pages 44 & 45 Frightening forces and Forceful sports quiz
Learning objective
Considering different sources.
Using available sources.

Use photocopiable pages 44 and 45 as part of a class quiz set by the children themselves. The photocopiable sheets come complete with answers to give your class the feel for researching and setting these sorts of questions for themselves. Encourage your class to access the Internet and use school books to help them in their research. Use the questions in a class quiz with mixed-ability teams. Encourage the children to use the questions to compile a class quiz book and to try them out on other classes.

Page 46: Think like Newton!
Learning objective
Gravitational pull and opposing forces.
Balanced and unbalanced forces.

Use this quiz to encourage your class to see themselves as the scientists of the future. Although Isaac Newton lived many years ago, he too was once at primary school and the lessons he learned there, and most importantly the games he played there, had a large influence on his later discoveries. A good example of this can be seen in the activity 'The centrifugal effect' (photocopiable page 26).

Encourage the children to acknowledge each other's achievements, using photocopiable page 46 as the starting point for a larger class quiz, awarding 'Isaacs' as class awards for excellent effort.

Answers: 1 b) Newton actually invented the cat flap! He built the first one at the farmhouse where he grew up. When the cat had kittens he made a mini cat flap for them to use. **2 b)** The sides of the bubble bend light until it breaks into the colours that make it up. That's why you can see rainbows in bubbles. **3 b)** Newton just wanted to get rid of the woman but according to the story that's where the purse was found! **4 a) 5 b)** It's shocking ... but true! **6 a)** Newton had William Chaloner hanged until he was half-dead and then cut down to have his guts pulled out and his body chopped into pieces and his head cut off. (I expect he was 100 per cent dead after that.)

Page 47: Gruesome zero gravity
Learning objective
Gravitational pull and opposing forces.
Balanced and unbalanced forces.

Start this session by talking to your class about gravity and how it affects our everyday lives. Ask the children to imagine life without gravity, using the weightlessness of space as an example. Use photocopiable page 47 to focus your class on one aspect of life without gravity, expanding this in to an imagined report detailing a day with zero gravity.
Answers: First **b)**, second **e)**, third **d)**, fourth **c)**, fifth **f)**, sixth **a)**. Still wanna be an astronaut?

Page 48: Wordsearch
Learning objective
Considering different sources.
Using available sources.

Use this activity as a starting point, encouraging the children to compile their own wordsearches based on personal research about forces, trying them out on their partners.

MAY THE FORCES BE WITH YOU

NAME _____   DATE _____

Introducing Isaac

One moonlit evening, Isaac Newton was sitting in the orchard trying to figure out how the moon went round the Earth.

'Isaac! Your supper's on the table and it's your favourite!'

'Coming, mother!'

Isaac sighed and reluctantly closed his book. There was a silent snap. The slender stalk holding the apple to the tree gave way. Wrenched by an unseen force the apple hurtled downward. It tumbled through the rustling leaves and bounced on Isaac's brainy bonce.

He rubbed his head and looked at the moon. It shone like a bright silver coin in the evening sky. 'So why doesn't the moon fall too?' he asked himself.

For some strange reason Isaac remembered his school and the dreaded 'bucket game'. He remembered having to whirl a bucket of water around his head on a rope. It was hard work and Isaac was a thin little boy. But amazingly all the water had stayed in the bucket as if trapped by an unseen force.

'Maybe that's what keeps the moon in place,' he murmured.

Isaac had forgotten his supper. He was calculating how strong gravity would need to be to stop the apple sailing into space. Then he thought about how fast the moon has to move to prevent it crashing down to Earth.

● Look around you. What examples of gravity can you see? Give three examples below.

1	2	3

● Try these true or false questions and research your answers.

1 There is a unit of measurement named after Isaac, called the 'Newton'. TRUE/FALSE
2 Isaac Newton was married six times and had thirteen children. TRUE/FALSE
3 Isaac Newton developed three rules about forces that we still use today. TRUE/FALSE
4 Isaac Newton once had a man put to death for forging money. TRUE/FALSE

● Now try out some of your own true or false questions on your partner.

NAME _____ DATE _____

Newton's First Law

- Newton's First Law says that if left alone, a moving object carries on moving in a straight line until another force makes it change direction.

- Put Newton's First Law to the test! You will need a partner, a teaspoon, ten counters and the target below.

- Place the counters on the end of the spoon handle and flick them at the target below.

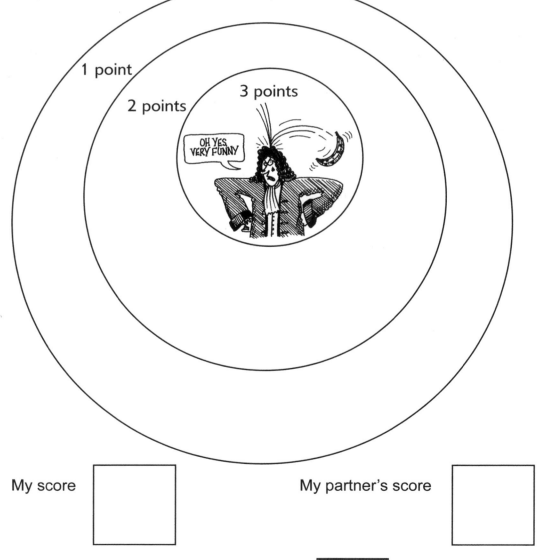

My score [ ]   My partner's score [ ]

NAME _____ DATE _____

NEWTON'S SECOND LAW

- Newton's Second Law says that when a force is applied to an object, it changes its momentum. It moves.

- So a strong kick will send a football further than a little tap.

- To play flick football you will need a partner, table tennis ball, pipe cleaners and a clear desk.

- Mark out the pitch on A3 paper and use the pipe cleaners as goal posts.

- The idea is to flick the ball from your half into your partner's goal in a set number of flicks. Put a tick if you score, a cross if you miss. Take it in turns.

Number of flicks	1	2	3	4	5
Me					
My partner					

- Can you score with just one flick?

- Take three penalties each. Use a strong flick for one and a weaker flick for the others.

  What do you notice?

_____

- Why do you think this is?

_____

_____

- Compare your scores and answers with other groups.

NAME _____   DATE _____

Newton's Third Law

- Newton's Third Law says that when an object exerts a force onto a second object then the second object will push back just as hard.

- So if you walk into a lamp post it pushes back with equal force!

- Try this simple experiment to see how Newton's Third Law works. You will need Plasticine, a selection of light and heavy books and a cookie cutter.

- Flatten the Plasticine so that it is 4cm thick. Place the cookie cutter on it without putting any pressure on it.

- Place an exercise book on the cookie cutter and leave it for ten seconds. What do you see when you remove the book and the cookie cutter?

- Now repeat the test using a text book. How thick should the Plasticine be for a fair test? Record your observations.

- Repeat once more, using the heavy book. How long should you leave the book on the cutter for a fair test? Record your observations.

Draw your observations here

NAME _____   DATE _____

NEWTONS

● Isaac Newton was a scientist who researched forces.

● A unit of measurement was named after him.

● Choose a selection of everyday objects and estimate their weight in Newtons. Now weigh these objects using a Newton meter. Work out the difference between your estimate and the actual weight after each one.

● Show any working out or calculations.

Bet you never knew!
When Newton's apple hit the ground the Earth bumped against the apple. That's what Newton's Third Law says: things always push back with equal force. But the Earth moved such a tiny distance no one noticed. Oddly enough a unit of force was later named the 'Newton' in the scientist's honour. And the weight produced by one Newton is roughly the same as … an apple.

Object	My estimate (N)	Actual weight (N)	Difference (N)

Newton's dog, Diamond, knocked over a candle and caused a fire that destroyed years of Newton's work on light. What did the scientist say?

a) SAY YOUR PRAYERS, MUTT – YOU'RE HISTORY!

b) THAT'S ALRIGHT OLD BOY – ACCIDENTS HAPPEN!

c) O DIAMOND, DIAMOND, YOU LITTLE KNOW THE DAMAGE YOU HAVE DONE!

NAME _____ DATE _____

Measuring on the moon!

- The force of gravity on the moon is only a sixth of what we experience here on Earth. This means that everything weighs six times more here than on the moon!

- Take a collection of objects and weigh them with a Newton meter.

- Record their weights in Newtons and then calculate what their weights would be if you repeated your experiment on the moon.

- Work out the difference.

Object	Weight on Earth (N)	Calculated weight on the moon (N)	Difference

NAME: Gravity

THE BASIC FACTS: You'll find gravity between any two things. The larger one somehow tugs at the smaller. This effect is usually slight and you won't notice it unless the big object is MASSIVE. Scientists think objects do this by producing tiny invisible bits called gravitons that carry the force.

THE HORRIBLE DETAILS: The force of gravity has been used for some horrible executions. In fact, whenever you fall, gravity brings you down to Earth with a bump.

SMALL THING

BIG THING (THE EARTH)

Idle inertia

Physicists use the word inertia to describe how things stay the same. Motionless things stay idle and moving things carry on until another force gets in the way. That's Newton's First Law again.

Dare you discover … the inertia of an egg?

What you need:
A plate
A raw egg
A hard-boiled egg

What you do:
1 Gently spin the raw egg on the plate.
2 To stop the egg touch it with your finger.
3 Gently lift your finger up.
4 Now repeat steps 1–3 with the hard-boiled egg.

What do you notice?
a) When you lift your finger the hard-boiled egg continues to spin.
b) When you lift your finger the raw egg continues to spin.
c) When you lift your finger the raw egg spins and the hard-boiled egg rocks from end-to-end.

DON'T PUSH DOWN TOO HARD

WHOOPS

Important note: The egg should spin on the plate. Not spin through the air and smash on the floor. If this happens you'll be force-fed omelette!

● Write up a full report that includes your results and observations.

● Include your conclusions at the very end.

MOMENTUM

- Momentum is all around us, in fact it keeps us going!

- Take a look at this fact file.

NAME: Momentum

THE BASIC FACTS: Momentum keeps you moving. That way you don't smash Newton's First Law. (That's the one about going in a straight line unless something stops you.)

THE HORRIBLE DETAILS: Momentum makes your stomach jump when you go over the top on a roller coaster.

ARGHHH!

The momentum of your half-digested food carries on up. If it comes up too far it could be fatally embarrassing!

Murderous momentum facts

1 In 1871, showman John Holtum tried to catch a flying cannon-ball with his bare hands. It wasn't fired from a real cannon, of course. Holtum used a specially built gun that fired a slow-moving ball. But even so he nearly lost a finger. The stunt proved very popular and John bravely practised until he'd perfected the trick. He should have changed his name to 'Halt-em'.

2 In nineteenth-century America railways were rarely fenced off and brainless buffalo often blundered onto the tracks. To tackle this menace, by the 1860s trains were fitted with wedge-shaped 'cow catchers'. The idea was that the train's momentum would scoop the buffalo out of harm's way.

GEE, SORRY MA'AM!

CLONK

3 In Finland, elk (otherwise known as moose) cause fatal road accidents. When hit by a car, the momentum of the car flips the moose over. So the loose moose lands on the car roof. Its weight crushes both the car and its driver. Perhaps the cars should be fitted with 'moose catchers'.

- Add some examples of momentum in your classroom.

NAME _____   DATE _____

Galileo's gravity 1

- Galileo was an Italian scientist who lived nearly 500 years ago.

- He performed some very imaginative experiments to prove his ideas about measuring forces.

- Try out Galileo's experiment for yourself.

- You will need a beach ball and a leather football of the same size.

- Record the weight of both objects.

- Take them to a first-floor window. Check that no one is underneath! If there isn't a first floor, you can stand on a chair.

- Drop the objects at the same time.

GALILEO'S BOOK of EXPERIMENTS

People laugh at me when I say that light and heavy things fall at the same speed. They say "Of course heavy things fall quicker - 'cos they're heavier." Grrr! I'll show them.

Experiment 1

1 Climb leaning Tower of Pisa with two balls of the same size. One ball is wooden and one is metal. Make sure the metal one is much heavier.

2 Reach top of tower. It's really slippery and there's no hand-rail. Careful now!

3 Chuck the two balls off the tower. Try not to chuck self off at same time.

4 Oh - nearly forgot. Check no one's underneath.

5 Note how balls land. If I'm right they'll both land at the same time.

WOOD → ● ○ ← METAL

MIAOW!

What do you think will happen?

What happens?

Was Galileo right with his hypothesis?

NAME _____   DATE _____

Galileo's gravity 2

- Galileo was an Italian scientist. He worked for many years trying to prove his theories about forces.

- Try this experiment yourself.

  1 You will need a marble, a table tennis ball and a bouncy ball.

  2 Weigh each of these objects and record this data.

  3 You will also need a very long, wide tube, some books and a stopwatch.

  4 Set the tube so that it is resting against the books at an angle. Roll each object down the tube, timing how long it takes to get to the other end.

- What do you think will happen?

  _____

  _____

  _____

- What actually happened?

  _____

  _____

  _____

- Was Galileo right with his hypothesis?

  _____

  _____

GALILEO'S BOOK of EXPERIMENTS

People still don't believe me. Huh - this'll teach them a lesson.
ME

Experiment 2

1 Get a wooden board with a little wooden gully on it. Line it with some nice shiny parchment made from animal skin with the fat scraped off.

USE SKIN FROM CAT KILLED IN EXPERIMENT 1

2 Raise the gully on a slope and roll a bronze ball down it. (If you don't have a bronze ball any other metal will do.)

3 Be sure to precisely measure the time taken for the ball to roll to the bottom of the slope. OOPS - silly me, I was about to forget, no one's invented an accurate clock yet. Better use pulse to time ball's speed. Mustn't get too excited, or my pulse will be racing. Better repeat the test a few times to make sure.

3, 4, 5, 6, 7...

RUMBLE RUMBLE

4 I believe gravity makes things accelerate at the same speed. If I'm right balls of different weights will roll at the same speed too.

NAME _____ DATE _____

Friction

- Rub your hands together very quickly for ten seconds.

- What do you notice?

- This is the result of a force we call 'friction'.

- Try this friction experiment:

1 You will need a stopwatch, a marble and a protractor.

2 You will also need four pieces of cardboard, each 1 metre long. Cover one with sandpaper, the second with bubblewrap, the third with silky material and leave the fourth plain.

3 Set each piece of cardboard against a pile of books at 45 degrees, just like a slide.

NAME: Friction

THE BASIC FACTS: You get friction when two moving objects brush together. Tiny bumps on each side stick together. They make heat and sound as the energy of moving objects turns into heat and sound energy.

THE HORRIBLE DETAILS: Friction causes problems for machines because it slows them down or makes them overheat. But lack of friction also causes fatal problems. If your bike brake blocks get worn they can't grip the wheels with enough friction. So you can't stop. Help!

4 Roll the same marble down each one and time how long it takes to reach the bottom. Observe how it moves and any differences that you can see.

- How are the surfaces affecting the time and movement of the marble?

_____

- Which surface produced the most friction?

_____

- Which produced the least?

_____

- Why do you think this is?

_____

- Explain how you kept your test fair.

_____

AVOIDING FRICTION!

Sometimes we want friction. Brakes, tyres, rubber-soled shoes, sandpaper and driving belts in machines would be useless without it.

THESE 'FRICTION' BOOTS ARE GREAT FOR WALKING UP STEEP HILLS!

But sometimes we don't want friction. We want things to go smoothly. That's why some slippery character invented lubrication. A lubricant such as oil fills out the little bumps that cause friction and allows the surfaces to slide past one another.

Most winter sports depend on lubrication. Sledges, skis and skates move easily because they melt a thin layer of ice beneath them. So they float along on this watery lubricant without too much friction. Until you slip over.

LOTS OF FRICTION

VERY LITTLE FRICTION

Dare you discover ... how to give things the slip?
What you need:

BANANA

COOKING OIL

PLASTIC BOTTLE TOP

KITCHEN TOWELS

TWO PLASTIC FOOD TRAYS

What you do:
1 Flick the bottle top along the first tray. Make sure the top stays on the tray and doesn't fly through the air.
2 Carefully pour a few drops of cooking oil on the first tray. Smear it over the surface with a kitchen towel until the surface is shiny and there is no extra oil on the tray.
3 Now flick the bottle top again as hard as before. Note what happens.
4 Mash up the banana and, using another kitchen towel, smear a little of the mixture over the second tray. Make sure the surface is smooth and shiny and there are no lumps of banana left.
5 Now flick the bottle top again as hard as before.

What do you notice?
a) Both the oil and the banana make good lubricants. They help the top move faster.
b) The top stuck to the banana and skimmed along over the oil.
c) The top stuck in the oil but skimmed over the banana.

STRETCHING

● Dare you discover ...

... What happens when something stretches?

What you need:
Yourself
A 0.5 cm-thick elastic band

What you do:
Suddenly stretch the elastic band.
Put it against your face.
What happens and why?
a) The elastic band feels strangely cold because all the energy has been stretched out of it.
b) The elastic band feels warm. This is due to the energy that you have provided by stretching it.
c) The elastic band feels warm because stretching causes friction with your hot sweaty little fingers.

... The power of an elastic band?

Here's a machine that uses stored energy in an elastic band to get moving. Ask an adult to help with some of the cutting.

What you need:

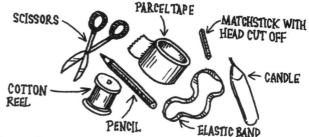

SCISSORS
PARCEL TAPE
MATCHSTICK WITH HEAD CUT OFF
CANDLE
COTTON REEL
PENCIL
ELASTIC BAND

What you do:
1 Cut 2.5 cm (one inch) off the bottom of the candle.
2 Remove the wick from the wax and make its middle hole large enough for the elastic band.
3 Pass the elastic band through the centres of the candle stump and the cotton reel.
4 Pass the matchstick through the elastic band at its cotton reel end. Secure the matchstick with a strip of parcel tape.
5 Pass the pencil through the elastic band at its candle end.
6 Wind the elastic band by turning the pencil. Watch your vehicle creep along as the elastic band unwinds. Compare its performance on rough and smooth slopes.

What do you notice?
a) It climbs better on smooth slopes.
b) It climbs better on rough slopes.
c) It can't climb slopes.

NAME _____ DATE _____

Springs around us!

1 The first toasters sold in 1919 had powerful springs that shot toast into the air. Bet that surprised a few people.

2 Springs sometimes break. Metal fatigue does for a cheap spring after about 100,000 extensions, but a better spring lasts over 10,000,000 extensions. A surprisingly long stretch.

3 You know the circus act where a person is fired from a cannon? You may be surprised to discover that springs rather than explosions are used to provide the necessary force. The bang is a firework let off to make it look like the cannon had really fired.

4 And did you know we've got springs in our legs too? The ligaments that hold your joints together are a bit springy and your 'S' shaped backbone jogs up and down as you walk. Together they'll put a spring in your step.

5 In the 1970s two American scientists trained a pair of kangaroos to hop on a treadmill. The scientists found that kangaroos jump using their springy tendons. It's a bit like jumping on a pogo stick.

6 Springy things are important for sport. Traditional tennis rackets were very expensive and strung with springy sheep's guts. Sounds like a bit of a racket. And talking about springy sports equipment, trainers have to be springy too.

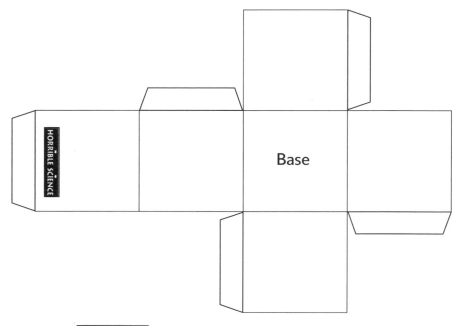

● Use the net to the right to construct your 'Jack in the Box'.

● Cut two long strips of paper and plait them into a concertina. Draw your own 'Jack' and attach it to the end.

● Stick Jack inside the box and close the lid.

● What happens when you open it?

NAME _____ DATE _____

Centrifugal effect and centripetal force

- These forces are all around us.

- A 'bolas' uses both these forces and is used for catching animals or people.

- You whirl the bolas above your head and let go. The rope winds around your target's legs!

O.K. – SO I NEED TO PRACTICE A BIT MORE...

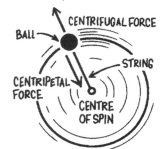

NAME: Centrifugal and centripetal forces

THE BASIC FACTS: Imagine whirling a small ball round your head on a bit of string.

CENTRIFUGAL FORCE / BALL / STRING / CENTRIPETAL FORCE / CENTRE OF SPIN

1. Centrifugal force tries to throw the ball off in a straight line.

2. Centripetal force is the opposite. It tries to pull the ball inwards towards the centre of its spin.

Dare you discover ... how a bolas works?

What you need:
Two balls of Blu-Tack each 2.5 cm (1 inch) across
A piece of strong string or twine 52 cm (20.5 inches) long

What you do:
1 Wrap a ball of Blu-Tack around each end of the string.
2 Squeeze the Blu-Tack to make sure it is holding the string securely.
3 Now you can practise throwing it. Hold the string between your thumb and fingers half-way between the two balls. Whirl the string round your head. Let go.

WHIZzzz

From your observations how does the bolas work?

a) Centripetal force makes the bolas fly off in a straight line. Centrifugal force helps the bolas wrap round the chair.
b) Centrifugal force makes the bolas fly off in a straight line. Centripetal force wraps the bolas round the chair.
c) Centrifugal force makes the bolas fly away at first, but centripetal force makes it come back like a boomerang.

HORRIBLE HEALTH WARNING!

1 Practising throwing your bolas indoors could be fatal for you if it knocks any priceless ornaments off the mantelpiece. Much better to practise outside in a wide open space.

2 Try to resist the temptation to throw your bolas at a small brother/sister/cat/dog, even if it is in the interest of science. You could use a small tree for target practice instead.

THE CENTRIFUGAL EFFECT

- This is your opportunity to carry out one of the most important Forces experiments of all time!

- You will need:

  A bucket

  Water

  Empty playground

- Isaac Newton hated this game when he was at school, but later it helped him work out how the centrifugal effect works!

- What do you notice?

- Write it up in your own words.

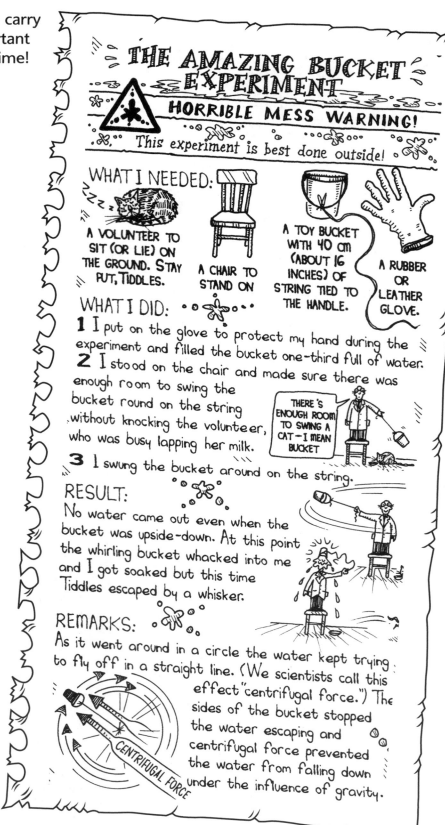

THE AMAZING BUCKET EXPERIMENT

HORRIBLE MESS WARNING!

This experiment is best done outside!

WHAT I NEEDED:

A VOLUNTEER TO SIT (OR LIE) ON THE GROUND. STAY PUT, TIDDLES.

A CHAIR TO STAND ON

A TOY BUCKET WITH 40 cm (ABOUT 16 INCHES) OF STRING TIED TO THE HANDLE.

A RUBBER OR LEATHER GLOVE.

WHAT I DID:

1 I put on the glove to protect my hand during the experiment and filled the bucket one-third full of water.

2 I stood on the chair and made sure there was enough room to swing the bucket round on the string without knocking the volunteer, who was busy lapping her milk.

THERE'S ENOUGH ROOM TO SWING A CAT — I MEAN BUCKET

3 I swung the bucket around on the string.

RESULT:

No water came out even when the bucket was upside-down. At this point the whirling bucket whacked into me and I got soaked but this time Tiddles escaped by a whisker.

REMARKS:

As it went around in a circle the water kept trying to fly off in a straight line. (We scientists call this effect "centrifugal force.") The sides of the bucket stopped the water escaping and centrifugal force prevented the water from falling down under the influence of gravity.

CENTRIFUGAL FORCE

NAME _____   DATE _____

favourite toy of Nobel prize winner Wolfgang Pauli (1900–1958) who was trying to work out the physics of inertia. Here's some crucial info to get you 'tops' of the class.

Tops balance because angular momentum keeps them going. They keep turning in the same way despite the efforts of gravity to pull them down. Bigger tops need more effort to get going but spin for longer. Tops are popular with kids the world over. Here's a traditional Inuit game you might like to play when it gets really cold.

IGLOO   SPINNING TOP

What you need:
An igloo
A spinning top

Spin the top. Run round your igloo (or house). Try to get back inside before the top falls down. (This could be fatal if you don't wrap up warm first.)

YOU ONLY HAVE TO GO ROUND ONCE!

Spinning

● Try the Igloo game, using a chair instead of an igloo.

● Can you make the top spin for long enough?

● Try making a walled arena, 30cm x 30cm. Spin two tops inside it at once, with a partner.

● What do you think will happen?

● What actually happens?

Top tricks
Physicists like nothing better than playing with their favourite toys. Well, according to them they're investigating forces. Oh, yeah.

CAN I HAVE A GO NOW?

LOOK, LOOK, I'VE DONE A REALLY LONG ONE!

FIRST PERSON TO DROP IT IS OUT

There are loads of toys that use the forces of spinning. Toys like yo-yos, hula-hoops, frisbees. And tops. A top was a

NAME _____   DATE _____

Push, pull and twist

● Pushing, pulling or twisting objects all involve forces.

● Which force is Sir Isaac Newton using on his quill?

Push, pull or twist?

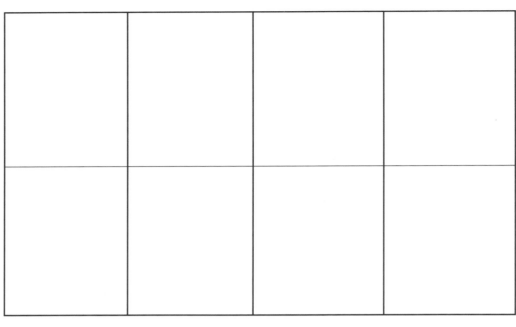

● Which force are the sheep using to graze?

Push, pull or twist?

● Which force is the cyclist using to turn the handlebars?

Push, pull or twist?

● Now find your own examples.

● Draw them in these boxes, cut them out and shuffle to play 'Push, pull or twist' Snap with your partner.

NAME _____   DATE _____

Air and water

Dare you discover … how air pressure helps you drink?

What you need:
Yourself
A bottle of your favourite drink (it's all in the interests of science) – just so long as the bottle's got a narrow neck.

What you do:
1 Try drinking from the bottle. Sit upright and tip the bottle up so it's level with your mouth. You can easily suck the liquid up.
2 Now stick the mouth of the bottle in your mouth. Wrap your lips around the neck of the bottle. Now try to drink.

What do you notice?
a) It's as easy as before.
b) You can't suck any more drink up.
c) You dribble uncontrollably into your drink.

NAME: Air and water pressure

THE BASIC FACTS: When tiny bits of air and water (molecules) are pushed aside by an object they push back. That's why when you get into a deep bath you can feel the water pushing against your body. It's what is called water pressure.

THE HORRIBLE DETAILS: The deeper you go the more water there is above you. This means more pressure. Divers breathe air that's also under pressure to stop their lungs getting squashed.

THIS SUBMARINE'S AMAZING – IT EVEN HAS A SHOWER IN IT...

ACTUALLY, THAT'S CALLED A LEAK CAPTAIN

((SUCK))
SUCK

Bet you never knew!
A vacuum is a completely empty space where there's no air or water pressure. Outer space is a vacuum and if an astronaut went into space without a space suit to protect them the air inside their body would explode and their eyeballs would plop out of their sockets. Erk!

● Take a look around your classroom. Make a list of everything that you can see.

● Imagine that all the furniture and school equipment has been taken out as well as the light bulbs, door handles and carpet.

● Would your classroom be empty or full? Explain your answer.

NAME _____   DATE _____

AIR PRESSURE 1

● Carry out this experiment, following the instructions carefully.

THE IMPOSSIBLE BAG TEST

WHAT I NEEDED:

AN EMPTY SWING BIN
(NOTE: IT'S QUITE ALL RIGHT
TO USE A NEW PEDAL-BIN BAG
AND AN EMPTY PEDAL-BIN.)

A NEW SWING
BIN LINER

WHAT I DID:

1 I placed the bag inside the bin with its edges hanging outside below the rim of the bin.

2 Now for the scientifically important bit — I tried to lift the bag out of the bin.

RESULT:

It's impossible without tearing the bag! As soon as I managed to lift part of the bag another part of it was sucked in! Tiddles tried to help and punctured the bag with her claw — it was then easy to lift out.

ME-OW!

REMARKS:

Air pressure squashes on our bodies with the weight of two elephants! Yes, Tiddles, and it squashes on you with the weight of an overweight horse!

AIR
PRESSURE

AIR
PRESSURE
SEALS
BAG
AGAINST
BIN
SIDES

The bag stopped air getting into the bin and the weight of the air pushing down on the bag kept it in place. Once the bag was holed, air could get underneath it and the bag was easy to remove.

Bet you never knew!
Imagine all those kilometres of air above you pressing down on your head. The air pressure on your body is an incredible 100,000 Pascals. That's the same weight as two elephants. Luckily, the air inside your body is under pressure too. It pushes outwards with the same force so you don't even notice it. Planes that fly at high altitudes have pressurized cabins in which the air is kept at the same pressure as ground level. If a pilot flew without his protection the lower air pressure would cause air bubbles in his or her body to get bigger.

● What do you think will happen?

_____

_____

_____

● Write up what actually happened in your own words.

_____

_____

_____

_____

_____

_____

_____

_____

_____

_____

_____

_____

NAME _____   DATE _____

AIR PRESSURE 2

● Carry out this experiment, following the instructions carefully.

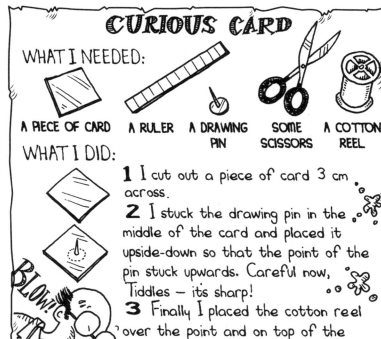

CURIOUS CARD

WHAT I NEEDED:

A PIECE OF CARD | A RULER | A DRAWING PIN | SOME SCISSORS | A COTTON REEL

WHAT I DID:

1 I cut out a piece of card 3 cm across.

2 I stuck the drawing pin in the middle of the card and placed it upside-down so that the point of the pin stuck upwards. Careful now, Tiddles — it's sharp!

3 Finally I placed the cotton reel over the point and on top of the card. I took a deep breath and blew down the hole while gently lifting the reel.

BLOW!

● What do you think will happen?

_____

_____

_____

_____

_____

_____

● What actually happened?

_____

_____

_____

_____

_____

_____

_____

● Write it up in your own words.

_____

_____

_____

_____

_____

_____

_____

_____

_____

_____

RESULT:

You might think that the air will blow the card downwards — but the pin and the card actually rose up with the reel. Well, it did until Tiddles knocked it with her paw and it fell off. Yes, Tiddles, I'll give you your supper in a moment!

HURRY UP!

REMARKS:

The air from my breath passed through the reel and spread over the card. Because the air in my breath was moving faster than the air under the card, it was pressing less hard (we scientists call this force air pressure). This allowed the air under the card to lift it up. This sounds a little complicated but hopefully my diagram will explain things better!

BREATH

AIR PRESSURE

NAME _____   DATE _____

Air resistance

- This is your opportunity to not only make a fully operational parachute but to find out how it works, all courtesy of air resistance!

- Cut out both rectangles.

- Attach 10cm of cotton to each corner marked with an 'X'.

- Join the threads together with tape.

- Attach paperclips to the thread.

- Drop both parachutes from the same height.

- Can you predict which one will be the first to land?

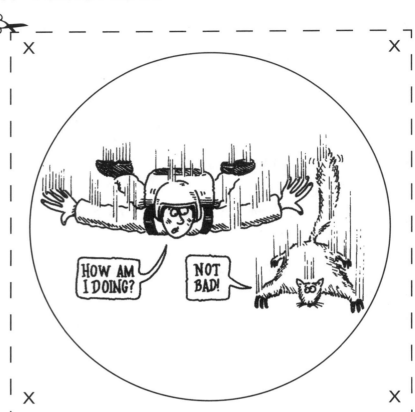

NAME _____   DATE _____

Air resistance report

● Use this report sheet to record your air resistance experiment.
● You can use words or drawings or both!

Equipment (what we used)

Method (what we did)

Hypothesis (what I think will happen)

Results (what actually happened)

How we made the test fair

Conclusion (what I learned)

NAME _____   DATE _____

Juggling

- You will need four newspaper balls, tightly rolled up – and plenty of space!

- If you have a mirror then you can use it to help, but it is not essential.

- Keep away from anything that may hurt you or break.

ADOPT A CONFIDENT AND RELAXED EXPRESSION

REMAIN CALM AND STILL

1 Stand in front of a mirror with your elbows tucked close to your body and your hands level with your waist. Place your legs apart with your knees slightly bent. Easy, isn't it? Are you ready?

2 Take a deep breath and let it out slowly. That's right – relax. Now without looking at your hands ... throw the ball gently up and over your head. Notice how it falls in an arc under the influence of gravity. Catch the ball in the palm of your other hand. Keep your eyes in the top part of the ball's flight. OK – that's the easy bit.

3 Now it gets a bit harder. Juggling with two balls takes a bit of practice. Throw one ball up as before. When the ball is just about to drop, throw your second ball up from the other hand. Ideally the second ball should pass just under the first ball.

4 OK, this takes practice. Better practise now to get it right.

5 This is where it gets really hard. Three balls. Sure you want to try? OK. Hold two balls in one hand and one in the other. Repeat Step 3.

6 Now here's the clever bit. When ball 2 is just about to drop throw ball 3 up and try to get it to pass under ball 2. Meanwhile catch ball 1 and throw it up just when ball 3 is about to drop. Easy!

7 Fantastic, keep going!

YIPEEE!

HERE GOES ...

BALL 3

NAME _____

DATE _____

TENNIS

The scientist's guide to tennis

Tennis ball seams are the same on each side. This means equal amounts of air turbulence. So the ball flies straight.

That's quite a velocity. Slice the racket downwards and you'll get back-spin. The ball tumbles backwards as it flies forwards. This drags air over it. As this air speeds up, the pressure above the ball drops and the greater air pressure under the ball raises it. We call this effect, lift.

DIDN'T THINK IT WOULD 'LIFT' THAT HIGH!

RACKET SLICES DOWNWARDS CREATING BACKSPIN ON BALL

BALL 'LIFTS' AS BACKSPIN CREATES LESS AIR PRESSURE ABOVE BALL AND GREATER PRESSURE BELOW

Top spin is the opposite. Strike the ball upwards and the ball tumbles forwards as it flies forwards. This drags air under the ball. And as it speeds up the pressure drops and the ball is pushed lower and it bounces faster.

DIDN'T THINK IT WOULD BOUNCE THAT FAST!

RACKET MOVES UPWARDS CREATING TOP SPIN

BALL BOUNCES FASTER AS TOPSPIN CREATES MORE AIR PRESSURE ABOVE THE BALL AND LESS BELOW

If you hit the ball a glancing blow it bounces extra slowly when it hits the ground. So it's even easier to whack.

- Use your tennis racket to hit the ball as far as you can.

- Draw a cartoon strip of your partner doing the same thing, showing each action. Include the ball's journey, right up until it stops.

- Label your cartoon strip and use arrows to show the different forces at work.

1	2
3	4
5	6

NAME _____ DATE _____

Bouncing

- You will need a metre stick, table tennis ball, football, rubber 'bouncy ball', a ball made of tightly scrunched paper.

- Can you predict which balls are the most and least bouncy? Remember to give a reason for your predictions.

_____

_____

_____

_____

_____

- Drop each ball from the same height.

- Measure how high it bounces up.

- Include observations about how many times the ball bounces, and the height that it reaches each time until it stops.

NAME: Bouncing

THE BASIC FACTS: When a rubber ball hits the floor the springy coiled rubber molecules that make up the ball are all squashed together. They soak up the energy of the impact and then bounce out again – making the ball bounce.

THE HORRIBLE DETAILS: The first chewing gum was made of chicle, a type of tree sap. American scientists tried to make the chicle into a type of rubber but it wasn't bouncy enough. So they just chewed the problem over, or rather chewed the chicle.

Object	Bounce 1 (height in m and cm)	Bounce 2	Bounce 3	Bounce 4	Number of bounces in total

- How did you keep the test fair?

- Write your conclusions after you have finished the experiment.

NAME _____   DATE _____

VIBRATION

● Try these two experiments to discover just how much vibration is around us!

Have you ever watched a washing machine shuddering and shaking as it washes and spins the clothes. Perhaps you've bravely laid a finger on the machine and felt the shaking passing up your arms. These are vibrations. And beware. They can be vicious.

TRAPPING YOUR TIE COULD PROVE FATAL...

((WEIRD WHINE)) GLASSES

⚠ **HORRIBLE DIFFICULTY WARNING!**

This is a hard experiment so you may have to whine for adult help to get it right.

WHAT YOU NEED:
SOME ← WATER
A FINGER (YOUR OWN)
A WINE GLASS

WHAT YOU DO:

1 Wash your hands and dry them carefully. (I took care to wash my hands. The Professor might not have been amused if he found mucky smears over his best wine glasses.)

2 Moisten your fingertip (no, not with spit – use water!) and stroke it lightly around the top of the rim of the glass. Your finger should be just touching the glass but not rubbing or pressing against it.

STROKE!

Dare you discover ... how much your body vibrates?

What you need:

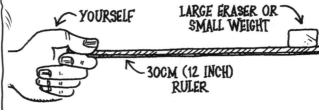

← YOURSELF

LARGE ERASER OR SMALL WEIGHT

30CM (12 INCH) RULER

What you do:
1 Place the eraser on the end of ruler.
2 Grip the opposite end of the ruler by your thumb and forefinger. Then hold the ruler as close to its end as you can.
3 Stretch out your hand balancing the eraser on the opposite end of the ruler.

What do you notice?
a) Nothing. I did the test for ten minutes and my hand was steady as a rock.
b) After a few seconds the end of the ruler began to dance around as my arm twitched.
c) I lost my balance and fell over forwards.

Weird whine glasses

● What do you think will happen?

_____

_____

_____

_____

● What actually happened?

_____

_____

_____

_____

_____

NAME _____   DATE _____

Testing balance

- Everything has a centre of gravity, even human beings.

- This tightrope walker is concentrating on the centre of gravity inside her body. As long as her centre of gravity is above the tightrope and her weights are evenly balanced on either side then she will not fall.

- Walk across your balancing bar. What do you do with your arms to help you?

- Why do you think it works?

- Try it again, this time holding a heavy book in one hand. Does the book make it easier to balance or harder?

- Try it again with a book of the same weight in each hand. What did you notice?

See if you can guess which of these incredible balancing acts are true and which are false.

1 In 1553 a Dutch acrobat balanced on one foot on the weathervane of St Paul's Cathedral, London waving a 4.6 metre streamer. And he didn't fall off.

2 In 1859 French tightrope walker Jean Blondin (1824–1897) walked across the raging Niagara Falls 50 metres in the air. And he was wearing a blindfold.
3 In 1773 Dutch acrobat Leopold van Trump juggled ten tomatoes whilst balancing on a tightrope 30 metres in the air. If he had fallen he might have invented tomato ketchup.
4 In 1842 a Miss Cooke wowed London circus goers when she sat at a table and drank a glass of wine. Boring? Not really. Everything was balanced on a high wire.
5 In 1995 Aleksandr Bendikov of Belarus balanced a pyramid made of 880 coins. The coin pyramid was upside-down and balanced on top of the edge of a single coin. Luckily, no one needed change for the bus.
6 In 1996 American Bryan Berg built a house of cards 100 storeys high – that's 5.85 metres.

7 In 1990 Brazilian Leandro Henrique Basseto cycled on one wheel of his bicycle for 100 minutes.

NAME _____   DATE _____

The flying ball test

● What do you think will happen? Write your prediction here.

_____

_____

_____

_____

_____

_____

_____

● What actually happened? Write your results here and check them against the answers.

_____

_____

_____

_____

_____

_____

_____

_____

THE FLYING BALL TEST

WHAT YOU NEED:
A HAIR-DRYER.
(I BORROWED ONE FROM MY COLLEAGUE WANDA WYE — WELL, I MAY HAVE FORGOTTEN TO ASK HER.)

A PLASTIC FUNNEL

A TABLE-TENNIS OR POLYSTYRENE BALL

HORRIBLE FAMILY WARNING!

Hair-dryers can be dangerous. **DO NOT** take the hair-dryer without asking. If it's your big sister's this could mean death. Oh, and don't put the hair-dryer anywhere near water.

WHAT YOU DO:
1 Set the hair-dryer to a low power setting and point it upwards. Place the ball in the stream of air. The ball floats. Tiddles! Leave that ball alone!

2 Now place the ball in the funnel and blow air up the spout.

What did you notice?

Answer: Look, Tiddles - the ball doesn't rise up! In the first case the force of the air pushing up from the hair-dryer was enough to support the weight of the ball. The funnel, however, makes the air rush around the ball rather than pushing the ball upwards. The slower-moving air above the ball presses it downwards, trapping it in the funnel. I have drawn a diagram to show the forces involved.

DON'T TOUCH THE FUNNEL WITH THE HAIR-DRYER.

BLOW!

AIR PRESSURE

NAME _____   DATE _____

AIR POWER

● What do you think will happen?
Write your prediction here.

_____
_____
_____
_____
_____
_____
_____
_____

● What actually happened?
Write your results here and check
them against the answers.

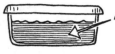

⁚AIR-POWER EXPERIMENT

WHAT YOU NEED: TWO GLASSES ⟶

A BATH OR WASHING UP BOWL
OF WATER. (I HAD TO DO THE
WASHING UP BEFORE I COULD
START THIS EXPERIMENT.)

WHAT YOU DO:

1 Hold one glass under water and turn it
upside-down. Then lift up the glass so that
the rim is almost clear of the water. The
glass is still full of water because air
pressing down on the water outside the
glass pushes the water upwards inside the
glass. Hmm, time for another diagram I think.

2 Put the second glass into
the water upside-down so
that air is trapped inside it.

3 Carefully move the second
glass so that it's just
beneath the water-filled
glass and slowly turn it the
right way up. An air bubble
should rise up into the
water-filled glass.

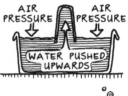
AIR PRESSURE   AIR PRESSURE
WATER PUSHED UPWARDS

AIR BUBBLE

_____
_____
_____
_____
_____
_____
_____
_____

Answer: The water-filled
glass empties! The force
of the air pushes the
water out. Yes, this
experiment is a real wash-out -
sorry, that was meant to be a joke!

EMPTY GLASS

NAME _____   DATE _____

Water wonder

● What do you think will happen? Write your prediction here.

_____

_____

_____

_____

_____

_____

_____

● What actually happened? Write your results here and check them against the answers.

_____

_____

_____

_____

_____

_____

_____

_____

_____

_____

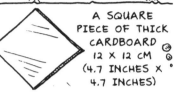

WATER WONDER

⚠ **HORRIBLE MESS WARNING!**

This experiment is best done outside. If you flood your house your pocket money might disappear down the plug hole!

WHAT YOU NEED:

A GLASS (BUT NOT A TALL ONE).

A SQUARE PIECE OF THICK CARDBOARD 12 X 12 CM (4.7 INCHES X 4.7 INCHES)

WHAT YOU DO:

1 Fill the glass so full of water that the water bulges over at the brim.

2 Gently push the cardboard down over the glass - try to make sure there are no bubbles and no water gets spilt.

3 Carefully turn the glass and the cardboard upside-down. You should be holding the glass with one hand and supporting the cardboard with the other. Once again you need to make sure that there are no bubbles and no water gets spilt.

4 OK - are you ready for this? Remove the hand supporting the cardboard... Yes, you did read that right. Well, get on with it! Everything's going to be fine ...

⚠ **HORRIBLE DIFFICULTY WARNING!**

You might need to practise this experiment once or twice to get it right. But don't practise this experiment with your brother or sister underneath.

Answer: The cardboard stays in place and the water remains in the glass! Since there is no air in the glass to push against the cardboard from above, the air pressure pushing up against the cardboard is strong enough to keep the cardboard pressed upwards against the rim of the glass.

AIR PRESSURE

NAME _____   DATE _____

PAPER POWER

● What do you think will happen?
Write your prediction here.

_____

_____

_____

_____

_____

_____

_____

PAPER POWER

WHAT YOU NEED:

A SHEET OF A4 PAPER

TWO LARGE TINS OR TWO PILES OF BOOKS

SOME SMALL WEIGHTS. (YOU COULD USE SMALL UNOPENED YOGHURT POTS OR TOY BRICKS.)

WHAT YOU DO:

1 Fold the paper into a zigzag shape with gaps of 2 cm between the creases.

2 Place the tins 15 cm between the creases. the paper on top of it so that the corrugated zigzag creases run between the two tins.

3 Now place your weights on the paper over the gap between the tins.

What did you notice?

● What actually happened? Write
your results here and check them
against the answers.

_____

_____

_____

_____

_____

_____

_____

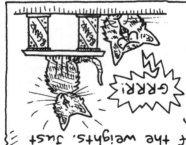

_____

Answer: Amazingly, the paper doesn't collapse! The zigzag shape of the paper spreads the force of the weights. Just one bit of paper can support up to 600 grams: that's the weight of an eight-week-old kitten and not a fat cat like you, Tiddles!

_____

_____

NAME _____   DATE _____

Body balance test

WHAT YOU NEED:

YOUR OWN BODY   A WALL   A COIN   A RULER

WHAT YOU DO:

1 Place the coin on the floor 70 cm from the wall.

2 Stand with your back to the wall and your heels touching the skirting-board.

3 Try to bend over and pick up the coin without your heels losing touch with the wall. (You're not allowed to crouch down either.)

WALL

COIN

70CM

NGHH!

● What do you think will happen? Write your prediction here.

_____

_____

_____

_____

_____

_____

_____

_____

_____

_____

_____

Answer: It's really hard! Your every move has to be balanced in terms of gravity. So when you bend you have to stick out your, er - what can I call it? Ah yes, your back-side to balance the force pulling on your upper body as you bend. With the wall in the way you can't do this - so it's almost impossible to pick up the coin without falling forward.

I WONDERED WHY - ARGH! NOW I KNOW!

● What actually happened? Write your results here and check them against the answers.

_____

_____

_____

_____

_____

_____

_____

_____

_____

_____

_____

_____

Answers (total score eleven points):

1 TRUE. When a train goes though a long tunnel it pushes air in front of it. In a large tunnel this can slow the train. The cross passages prevent this problem by allowing the air to escape.

2 TRUE. The airport is built on an artificial island and it's expected to sink under the weight of the buildings by 11–13 metres in 30 years.

3 TRUE. In industry, water-cutting uses fine high-pressure sprays mixed with tiny bits of grit.

4 FALSE. Gravity makes you shorter! Every night you grow around 8 mm because when you lie down gravity isn't pushing down on your spine. This means that you wake up taller in the morning and shrink during the day!

5 TRUE. If it bent any more it would snap.

6 FALSE. The air weighs 50,000,000,000,000,000,000 (50 million billion) tonnes, give or take a few grams, but that's only one-third the weight of the sea.

7 TRUE. But not so as you'd notice, otherwise windy days would last for ever as the Earth stopped turning in space.

8 TRUE. In bright sunny weather the air pressure is greater and this squeezes your bulging body and makes you feel leaner and fitter. In miserable overcast weather the opposite happens and your body feels bloated and squashy.

9 TRUE. In the 1970s two scientists monitored how much oxygen kangaroos needed to breathe and found they needed less than running humans.

10 FALSE. It's the opposite way round! As you travel east the spin of the Earth pulls on your body and slightly reduces your weight. This effect is also too small to notice.

11 TRUE. The faster something moves the harder it hits you. Fired like cannon balls the hard Edam cheeses were moving fast enough to kill several sailors and wreck the sails of the attacking ship, forcing it to retreat. Hard cheese to them, then.

Fold here

Frightening forces quiz

Forces affect the movement of an object – but which of these are forces facts and which are forces fiction? Answer TRUE or FALSE.

1 Long train tunnels are built in pairs and linked by cross passages to allow air to escape from the tunnel.

2 New Kansai International airport in Japan is designed to sink.

3 Water can be used to cut metal, rock, leather and paper.

4 The force of gravity makes you taller.

GRAVITY! GRAVITY! GRAVITY!
GRAVITY! GRAVITY! GRAVITY!

5 In a storm, a loaded supertanker bends in the middle by up to 91 cm.

6 Weight is one measure of the gravity acting on an object. The air weighs more than the sea.

7 The force of the wind is enough to slow the spin of planet Earth.

8 Good weather makes you thinner.

9 Scientists have found that jumping around like a kangaroo requires less energy than running.

10 If you run west you'll weigh less than if you run east.

11 In the 1840s a ship from Uruguay, South America, fired balls of cheese to beat off an attacking ship.

WELL, THAT'S HARD CHEESE! THEY'LL BE CHEESED OFF IF IT HITS THEM

● How did you do? Now research some of your own true or false questions and test them on your partner.

NAME _____ DATE _____

Answers (total score five points):

1 c) Weightlifters first lift the bar up to chest level. When the weights on the ends of the springy steel bar spring upwards the movement helps the weight to be raised more easily.

2 b) The cars had to be fast. If they went slower than 177 km per hour they fell down the steep sides of the track under the influence of gravity.

3 a) Cyclists try to be as streamlined as possible to reduce drag.

4 d) It took seven dead sheep to make one racket. Can ewe believe it?

5 The missing word is AIR. Fancy a few hours moulding cow poo? By the way, the rules also state that the cow pats must be 100 per cent poo. So you need a breath of fresh air? You're just about to get one!

IT TAKES GUTS TO PLAY THIS GAME!

YOU FORGOT TO PUT THE PETROL CAP BACK ON

SPLOSH!

Fold here

Forceful sports quiz

Simply match the missing words to the sport. Just to make it a bit harder there's a question where we haven't given you the word. (You'll have to work this word out for yourself.)

Missing words
a) tights
b) steep
c) springy steel
d) sheep guts

1 Weightlifters use _____ to help them lift weights.

2 In the 1920s US racetracks had _____ sides to make the cars drive faster.

3 Cyclists wear _____ or shave their legs to increase their speed by up to 10 per cent.

4 Traditional tennis rackets got their springiness from _____

5 In the world cow-pat-hurling championships the rules say that cow pats can't be moulded to reduce drag. Drag is the _____ force of the _____ hitting the moving object.

● How did you do? Now research a missing word quiz of your own and test it on your partner.

Think like Newton!

1 YOU are Isaac Newton. You're doing your Science homework but the cat keeps wanting to come in and then go out again. What do you do?
a) Lock her in the shed.
b) Invent the cat flap.

2 Your neighbour spots you foaming at the mouth one day. In fact, you're blowing bubbles – but why?
a) You're playing with the cat.
b) You're studying the way the bubbles bend light.

3 After you become a famous scientist, some people think you have magical powers and a woman asks you to use magic to find her lost purse. What do you say?
a)

GET LOST, YOU SILLY OLD CRONE!

b) You say the magic word "ABRACADABRA" and send her off to look around the Royal Naval Hospital in Greenwich.

4 You fall out with your fellow scientist Robert Hooke after he claims your experiments on light don't work properly. What do you do?
a) Delay publishing your work on light until Hooke is dead so he can't have a go at you.

RIP
R. HOOKE
1635-1703

IT'S A GRAVE SITUATION ...YIPPEE!

b) Publish the work immediately and ask Hooke to make constructive comments.

5 You reckon your theory of gravity can predict the effects of gravity on an object to an accuracy of 0.00003 per cent. How do you persuade other scientists of this?
a) You spend years trying to prove the accuracy of your prediction.
b) You ask your publisher to print a fiddled set of figures that appear to show this result.

6 You take charge of Britain's coinage. Part of the job involves tracking down people who make fake coins, and by bribing informers and listening to gossip in shady pubs you eventually trap master forger William Chaloner. The penalty for forgery is death and the criminal begs for mercy. What do you do?
a) You arrange to have him executed in an especially cruel and disgusting fashion.

CENSORED!

ARGH, THAT'S FREEZING!
ERK, THAT'S BOILING!
OUCH, THAT'S SHARP!
N-N-NO THAT'S ARGH!

b) You ask the king to spare William's life.

NAME _____ DATE _____

Gruesome zero gravity

- Once astronauts are in space they experience weightlessness as they are no longer within reach of Earth's gravity.

- How would you cope?

Could YOU boldly go where not too many people have been before?

So how do you use a space toilet? Here are the stages involved, just put them in the right order. Any mistakes and you'll be covered in unmentionable substances. (You can have one point for each stage listed in the right order.)

WARNING: this quiz is a bit rude so don't leave it on your granny's chair!

a) Don't forget to wipe the toilet and your nether regions using a special wipe.
b) Sit down on the toilet seat.
c) Switch on the fan to suck the poo and pee out of the base of the toilet otherwise it will splurge all over you.
d) Grip the handles on either side to stop your body floating off the loo at a vital moment.
e) Connect your vital bits and pieces to the right-sized funnel or nozzle and … NO, I'm not going into details.
f) Press the button to increase the suction for poo.

- Write a report about your life as an astronaut.

- Give detailed descriptions about how you overcame the challenges of zero gravity!

NAME _____ DATE _____

WORDSEARCH

● Try this wordsearch. The words you are looking for are written in CAPITAL letters in the clues. They can be found backwards, forwards, sideways and diagonally in the grid below.

1 Sir Isaac NEWTON wrote three main Laws to explain forces around us.
2 GRAVITY became clear to him after an apple fell on his head!
3 FRICTION is a story that has been made up. True or False?
4 INERTIA is a word to describe how things stay the same.
5 The opposite of inertia is MOTION. True or False?
6 For every force that acts, another reacts – just like when you BOUNCE!

F	G	R	P	L	O	N	B
T	R	E	C	N	U	O	B
C	A	I	Z	E	M	O	M
H	V	A	C	W	P	Y	O
I	I	L	O	T	V	E	T
J	T	O	N	O	I	E	I
S	Y	Y	X	N	T	O	O
A	I	T	R	E	N	I	N

● Now try making your own wordsearch grid using forces for clues!